Astronomy in the 21st Century:

The James Webb Space Telescope; Facts and Objectives.

Anthony M. Acevedo

Table of Contents

Chapter 1: **What is the James Webb Space Telescope**.

The James Webb Space Telescope replaced the Next Generation Space Telescope as the designation of the space telescope in September 2002.

JWST stands for James Webb, a former NASA administrator. From 1961 until 1968, Webb served as the agency's director before stepping down only a few months before NASA sent a man to the moon.

Webb is regarded as a pioneer in space research, even though the Apollo lunar mission is most commonly connected with his time as NASA administrator. Webb established NASA's scientific goals even during a period of intense political unrest, arguing that the launch of a large space telescope should be a top priority for the space agency.

Under Webb's direction, NASA launched more than 75 space science missions. Missions that looked at the sun, stars, and galaxies as well as the area right above the Earth's atmosphere were among them.

The name chosen for the space telescope has not been well received by everyone. Due to allegations that Webb participated in discrimination against homosexual and lesbian NASA personnel while serving as an administrator, detractors started an online petition asking NASA to rename the telescope. Despite criticism, NASA has subsequently declared that it would not rename the telescope.

The James Webb Space Telescope (JWST) is on a mission to observe some of the faintest, oldest objects in the universe, from a vantage point nearly 1 million miles (1.5 million kilometers) from Earth. JWST launched Dec. 25, 2021, at 7:20 a.m. ET (12:20 p.m. GMT) from the Guiana Space

Centre (also known as Europe's Spaceport) in French Guiana.

The 21-foot, gold-coated main mirror of the Webb telescope was completely deployed on January 8, 2022, marking the successful conclusion of all significant spacecraft deployments in preparation for scientific operations.

Gregory L. Robinson, the Webb program director at NASA Headquarters, remarked that the accomplishment of all of the Webb Space Telescope's deployments was historic. "This is a great achievement for our team, NASA, and the globe," the project team said. "This is the first time a NASA-led mission has ever tried to complete a difficult sequence to unfurl an observatory in orbit."

The Webb space telescope was launched into orbit around the Sun on January 24, 2022, at the second Lagrange point, or L2, which is roughly a million miles away from Earth.

Bill Nelson, the NASA administrator, welcomed Webb back home. "Congratulations to the crew for their tireless efforts in making sure Webb arrived at L2 safely today. We've made progress in our quest to understand the secrets of the cosmos. And this summer, I can't wait to witness Webb's first fresh perspectives on the cosmos!

The positioning of Webb allows for a broad view of the universe and will protect the telescope's optics and research instruments.

The first full-color picture obtained by JWST, which scientists described as the deepest view of the cosmos ever recorded, was shared by President Joe Biden on July 11. The next day, NASA published four more first photographs to demonstrate Webb's extraordinary capability. These images included close-ups of a faraway dying star, an extraterrestrial exoplanet, and a cluster of five galaxies that were chaotically merging.

As the successor to the Hubble Space Telescope, a space observatory that is still in use and producing amazing photographs of the universe, Webb has a lot to live up to. Since Hubble's 1990 debut, it has shed a previously unheard-of amount of light on the universe's marvels. When it first started operating, research on cutting-edge subjects like dark energy and exoplanets was but a pipe dream. In addition, it has successfully captivated the public's attention to the point that it has become a household name.

The European Space Agency (ESA), the Canadian Space Agency (CSA), and NASA are partners in the operation and financing of the James Webb Space Telescope, sometimes known as Webb (like "Hubble"). According to NASA, the telescope is named after James E. Webb, one of the agency's first managers, who supervised the development of the Apollo program in the 1960s.

Nearly 20 years ago, in 2002, Webb's name was originally used to allude to what was then known as the "Next Generation Space Telescope." According to Live Science sister site Space.com, that choice was later questioned as the launch of JWST drew near. Many scientists claimed that Webb had discriminated against gay and lesbian NASA employees while serving as the agency's administrator and that as a result, his name shouldn't be attached to the well-known observatory. According to Space.com, NASA said in September 2021 that they will not rename the mission.

NASA's James Webb Space Telescope will launch from the ELA-3 Launch Zone of Europes Spaceport at the Guiana Space

Center in French Guiana on December 25, 2021, using an Arianespace Ariane 5 rocket.

The Atlantic said that Webb was once anticipated to cost half a billion dollars and be ready for launch in 2007.

Given the spacecraft's very complicated and cutting-edge design, these forecasts, however, proved to be too optimistic. According to the U.S. Government Accountability Office, the cost of building the telescope came to close to $10 billion, roughly tripling the previous estimate from 2009.

However, the scientists working on the research are certain that the outcomes would more than justify the time and money spent on it. NASA wants to make it clear that Webb is more than just a larger and more powerful telescope than Hubble. With more than 2.5 times the diameter and 100 times the sensitivity, the JWST is both of those things, but at its core, it is a very new kind of instrument.

A COMPARISON OF THE HUBBLE AND JAMES WEBB SPACE TELESCOPES

To progress science, one must "stand on the shoulders of giants," and the JWST will accomplish precisely that since the Hubble Space Telescope's findings served as the inspiration for its scientific objectives.

While Hubble mainly investigated the universe in the optical and ultraviolet spectrum, the two space telescopes have distinct capacities (with some infrared capabilities.) The JWST's primary viewing mode will be infrared. According to the ESA, redshifted light is light from distant objects that has shifted to longer wavelengths at the redder end of the spectrum as a result of the universe's expansion. The JWST will carefully study this infrared light and provide information about some of the earliest stars and galaxies in the cosmos.

The fact that JWST will circle the sun while Hubble will orbit the Earth is another significant distinction between the Hubble and James Webb space telescopes. Unlike Hubble, which was accessible and maintained by space shuttle flights, JWST will be too far away to be serviced.

Chapter 2: **Facts about James Webb Space Telescope.**

1. The Hubble Space Telescope was replaced by the James Webb Space Telescope, which is lighter. Most folks find this one to be rather shocking. In most cases, if you wish to construct a larger replica of anything, it will be more substantial and heavy. When compared:

Hubble had a 4.0 square meter collecting area, a 2.4-meter diameter main mirror, and a solid primary mirror.
James Webb has a collecting surface of 25.37 square meters, is 6.5 meters in diameter, and is constructed of 18 distinct mirror segments.

But if we were to weigh them both here on Earth, Webb would weigh 14,300 pounds with a mass of around 6,500 kilograms. With its improved instrumentation, Hubble now has a mass of 12,200 kg and a weight of 27,000 pounds, compared to 11,100 kg and 24,500 pounds when it was launched. As almost every component on James Webb is lighter than its Hubble counterpart, this is an amazing achievement of engineering.

2. The James Webb telescope's mirrors are the world's lightest big telescope mirrors. When the 18 main mirror segments are created, they are each shaped like a curved disk and weigh 250 kg (551 pounds). But by the time they're through, that bulk has been reduced to only 21 kg (44 pounds), a 92 percent weight loss.

It's remarkable how this is achieved. The mirrors are first chopped into their hexagonal form, which results in a minor bulk decrease. Then, however, and here is where it really shines, almost all of the bulk on the mirror's "back side" is machined away. What is left has undergone testing to guarantee that it will:

keep its exact form Despite being fragile, it can withstand launch stresses and vibrations as well as the projected frequency and speed of micrometeoroid hits.
be aware of the necessary form adjustments that the back-mounted actuators will make.
With the lightest such mirrors ever produced, these 18 mirrors will come together to create a single mirror-like plane with an accuracy of 18 to 20 nanometers, the greatest of all time.

3. *James Webb's mirrors are composed of beryllium, despite their apparent gold color.* Each mirror indeed has a gold covering, but it would have been disastrous to make the mirrors totally out of gold. No, not because of gold's very high density or malleability, both of which it unquestionably has. Thermal expansion would be the main issue.

Gold expands and contracts significantly with slight temperature changes, even at extremely low temperatures, making it impossible to use it as the primary component of Webb's mirrors. However, on this front, beryllium excels. You can guarantee that there will be defects in the mirrors at room temperature by chilling the beryllium to cryogenic temperatures and polishing it there. These imperfections will vanish once the mirrors are cooled back down to operational temperatures.

The gold coating is put on after the beryllium is produced and shaped to its final form.

4. The mirrors of the James Webb Space Telescope contain just 48 grams, or less than 2 ounces, of gold in total. The 18 mirrors used by James Webb must each be very good at reflecting the infrared light that it is intended to study. The appropriate quantity of gold must be used; if you use too much, the gold will start to expand, contract, and distort as the temperature changes. If you use too little, the mirror won't be completely covered.

Vacuum vapor deposition is the name of the procedure used to apply the gold coating. You inject a little quantity of gold vapor into the vacuum chamber after setting the "blank" mirrors inside of it. The back of the mirror and other areas that don't need to be coated are covered with masking tape so that just the polished, flat surface is covered with gold. This procedure is repeated until the gold is barely 100 nanometers thick, or around 600 gold atoms, as required.

The mirrors of the James Webb Space Telescope include a total of just 48 grams of gold, while the uninteresting rear sides have struts, actuators, and flexors connected to them.

5. A thin coating of amorphous silicon dioxide glass is applied to protect the gold from direct exposure to space. Why not just let the gold itself be exposed to the depths of space? Due to its extreme malleability and softness, it is very vulnerable to harm from even a little bump. A thin gold coating, in contrast to beryllium, would be significantly harmed by micrometeoroid impacts and hence be unable to maintain the smoothness required for the telescope's functioning without extra protection.

Amorphous silicon dioxide glass serves as the ultimate "coating above the coating" in this scenario. Although glass with a coating is normally used to create mirrors, in this instance the glass serves a very straightforward purpose: to remain transparent to light and safeguard the gold. Yes, it is covered with gold, but the gold itself also has to be protected by a coating.

6. James Webb's "telescope side" will passively cool itself to a temperature of no more than 50 K, which is cold enough to cause nitrogen to liquefy. Because it will be passively cooled in a manner that has never been done before, James Webb must be positioned near the L2 Lagrange point and not in low-Earth orbit like Hubble. James Webb has a massive, five-layered sunshield that has been specially designed to reflect as much sunlight as possible while protecting the layer below it. The infrared heat radiated by the Earth would prevent it from achieving the required low temperatures if it were in low-Earth orbit.

The sunshield itself is fashioned like a diamond and is 21.2 meters (69.5 feet) long by 14.2 meters (46.5 feet) wide. The "hot side" of each layer faces the Sun, while the "cold side" faces the telescope. On its hot side, the top layer will become as heated as 383 K, or 231 °F. The freezing side is down to 36 K, or -394 °F, while the hot side is still just 221 K, or -80 °F, by the time you reach the deepest layer. The telescope will be able to function as intended as long as it is kept below 50 K.

7. *Webb will reach a minimum temperature of 7 K with active cryogenic cooling.* All of Webb's near-infrared devices can operate at the low temperatures, in the 36 to 50 K range, that passive cooling achieves. This applies to three of its four main scientific equipment, the FGS/NIRISS (fine-guidance sensor/near-infrared imager and slitless spectrograph), NIRSpec (near-infrared spectrograph), and NIRCam (near-infrared camera). All of them are built to operate at 39 K, which is well within the range of passive cooling.

The cryocooler is used to cool the fourth instrument, MIRI (the mid-infrared imager) since it has to be chilled even more than passive cooling can. The MIRI device may be cooled to the required working temperature of 7 K by adding a liquid helium refrigerator to it since helium only turns liquid at around 4 K. The major consideration in most design choices for the James Webb Space Telescope was that the colder you can make your sensors, the longer the wavelength of light you want to explore.

8. *James Webb should continue to operate at its frigid temperatures for the duration of its life, unlike NASA's Spitzer, which changed to a "warm" mission when its coolant ran out.* Since it is a closed system, the liquid helium that keeps James Webb actively cooled should theoretically never run out. But no matter how thoroughly you protect against leaks, as everyone who has ever worked in experimental physics will confirm, they ultimately happen. Webb shouldn't run out of cryogenic coolant if it lives up to its design requirements as it was built for a 5.5-year mission at the very least, with the option of a decade or more in the best-case scenario.

It is always possible that something may go wrong and we won't be able to actively cool the mid-infrared imager effectively or for the duration of the mission, which would reduce Webb's sensitivity at increasingly longer and longer wavelengths. In the case of sunshield damage or inefficiency, the near-infrared sensors are subject to the same warning. The James Webb Space Telescope's wavelength range will become more restricted as it becomes warmer.

9.) *When it runs out of fuel, it will be condemned to a life-long sojourn in what is known as a "graveyard orbit" around the Sun.* More than three decades after its debut, Hubble is still in operation with help from four servicing missions. However, anytime Webb wants to accomplish anything that involves motion, it must spend its fuel. That contains: doing an orbital adjustment to maintain it in its L2 orbit performing a burn to correct its route towards its destination performing a burn to orient itself so that it points at its targeted target.

How much fuel we have left for scientific operations totally on how well the launch places Webb on its optimal trajectory toward its final destination. Fuel is limited.

Science procedures stop when the gasoline runs out. However, we can't simply let it float wherever it may go since it might jeopardize future L2-targeted operations. Instead, we'll send it to a graveyard orbit where it will orbit the Sun for as long as there is a Sun to orbit, just as we did for earlier spacecraft sent to L2, such as NASA's WMAP satellite.

10. Even though it wasn't intended to be maintained and upgraded, it might be robotically refueled to increase its lifespan. After all of this work, it seems unfortunate that Webb's lifespan will be so brief. Yes, five to ten years is plenty of time to learn a lot about the universe, accomplish a lot of ambitious scientific objectives, and prepare for unexpected discoveries that we may not have even yet envisioned. The fact that James Webb will have a lifespan that is cumulatively less than the complete amount of its tenure on Earth, however, seems inadequate given all we've gone through with development and delays.

Still, there is hope.

If we create the appropriate uncrewed technology, we might get access to a refueling port. The mission's lifespan might be extended by a decade or more with each refill if we can reach L2, dock with James Webb, access the refueling port, and refuel it. There have been reports that the German Aerospace Center, DLR, may be able to carry out precisely this kind of operation before Webb's end of life, which is most likely in the early 2030s. If Webb performs precisely as intended and is, as anticipated, fuel-limited, it could be the height of needless idiocy to not go with that choice.

Chapter 3: **Mission Objectives of the James Webb Space Telescope.**

Objective 1: The early universe

Webb is frequently characterized as a "time machine," which in a way it is. Because light from distant things travels with limited speed, we view them as they used to be in the past. Hubble has revealed us galaxies as they were many billions of years ago, but the JWST will be far more sensitive. NASA thinks it will see back to when the first galaxies formed, approximately 13.6 billion years ago.

And Webb has another edge over visible-band telescopes like Hubble.

Because the cosmos is expanding, light from distant objects is stretched out, increasing its wavelength. This implies light produced in the visible waveband reaches humans in the infrared, the band that the JWST is tuned for. One of its initial missions will be a survey, named COSMOS-Webb, of the most distant galaxies in a particular area of sky, to examine circumstances before the birth of the universe.

Objective 2: Galaxies throughout time

Thanks to Hubble's amazing images, most people know what galaxies look like: vast groupings of stars, typically grouped in perfectly symmetric spiral patterns. But they tend to be rather close galaxies, and consequently mature ones. The tantalizing peeks Hubble has offered of extremely early galaxies shows they are substantially smaller and messier-looking.

As yet, no one understands how these proto-galaxies originated, or how they subsequently clumped together to generate the bigger, regular-looking galaxies we see today, according to the California Institute of Technology(opens in new tab) (opens in new tab). It's believed that Webb will be able to address problems like these with its ultra-deep vision of the early cosmos.

Another well-established property of galaxies is the existence of supermassive black holes at the cores of most of them. In the early universe, these black holes typically powered extraordinarily light galactic nuclei called quasars, and Webb is slated to explore six of the most distant and brightest instances of these.

Objective 3: Lifecycle of stars
The galaxies that populate the cosmos formed very early on, and they've gradually developed ever since then. But that's not true of the stars within them, which go

through life cycles more analogous to living organisms. They're born, develop, age, and die, and the remains of ancient stars contribute to the raw material required to form new stars. Much of this process is widely known, but there's still a question surrounding the actual birth of stars, and the planetary discs that may develop around them.

That's because young stars are first encased behind a cocoon of dust, which regular telescopes utilizing visible light can't penetrate. But all this dust will be almost transparent at the infrared wavelengths utilized by Webb, so NASA expects it will finally expose the ultimate mysteries of star creation. In turn, this may tell us something about the beginnings of our sun and solar system.

Objective 4: Other worlds
One of the most fascinating areas of current astronomy is the hunt for exoplanets

circling other stars, especially Earth-like planets that may contain the chemical components and circumstances essential for life to emerge. The JWST will contribute to this quest in numerous ways, employing infrared imaging and spectroscopy to examine the chemical and physical aspects of planetary systems.

Its capacity to peek through dust and capture super-high quality photographs should offer us a direct glimpse of planetary systems — such as that of the freshly created star Beta Pictoris — in their very early phases, according to NASA's JWST website. Webb will also investigate the chemical makeup of exoplanet atmospheres, seeking in particular for tell-tale indications of the building blocks of life. This again is something an infrared telescope is perfectly equipped for since the chemicals building up planetary atmospheres tend to be most active at these wavelengths.